SPOTLIGHT ON OUR FUTURE

TECHNOLOGIES THAT HELP OUR WORLD

GENE BROOKS

NEW YORK

Published in 2022 by The Rosen Publishing Group, Inc.
29 East 21st Street, New York, NY 10010

Copyright © 2022 by The Rosen Publishing Group, Inc.

First Edition

All rights reserved. No part of this book may be reproduced in any form without permission in writing from the publisher, except by a reviewer.

Editor: Theresa Emminizer
Book Design: Michael Flynn

Photo Credits: Cover Andrey Armyagov/Shutterstock.com; (series background) jessicahyde/Shutterstock.com; p. 4 Maridav/Shutterstock.com; p. 5 Barcroft Media/Getty Images; p. 6 Nicole Patience/Shutterstock.com; p. 7 Bettmann/Getty Images; pp. 9, 13 (all) Bloomberg/Getty Images; pp. 10, 11, 15, 27 picture alliance/Getty Images; p. 12 oraziopuccio/Shutterstock.com; p. 16 anatoliy_gleb/Shutterstock.com; p. 17 Hannah Peters/Getty Images; p. 18 Alessandro Pierpaoli/Shutterstock.com; p. 19 monicaodo/Shutterstock.com; p. 20 Mohamed Abdulraheem/Shutterstock.com; p. 21 Ramin Talaie/Corbis Historical/Getty Images; p. 23 (all) Peter Parks/AFP/Getty Images; p. 24 Riccardo Mayer/Shutterstock.com; p. 25 John B. Carnett/Popular Science/Getty Images; p. 26 Lano Lan/Shutterstock.com; p. 29 SOPA Images/LightRocket/Getty Images.

Cataloging-in-Publication Data

Names: Brooks, Gene.
Title: Technologies that help our world / Gene Brooks.
Description: New York : PowerKids Press, 2022. | Series: Spotlight on our future | Includes glossary and index.
Identifiers: ISBN 9781725324350 (pbk.) | ISBN 9781725324381 (library bound) | ISBN 9781725324367 (6pack)
Subjects: LCSH: Technology--Juvenile literature.
Classification: LCC T48.B76 2022 | DDC 600--dc23

Manufactured in the United States of America

Some of the images in this book illustrate individuals who are models. The depictions do not imply actual situations or events.

CPSIA Compliance Information: Batch #CSPK22. For further information contact Rosen Publishing, New York, New York at 1-800-237-9932.

CONTENTS

HELPING HANDS 4
WHAT IS TECHNOLOGY? 6
WHAT IS CARBON CAPTURE? 8
NEW AND IMPROVED AIRPLANES 10
TECHNOLOGY AND FOOD 12
SOLAR ENERGY AND SOLAR GLASS 14
ENERGY STORAGE 16
POWER GRIDS 18
FROM PLANT TO PLASTIC 20
CLEAN OCEAN WATER 22
CLEAN DRINKING WATER 24
FARMING TECHNOLOGY 26
DRONES IN THE FUTURE 28
HELPING YOUR WORLD 30
GLOSSARY 31
INDEX 32
PRIMARY SOURCE LIST 32
WEBSITES 32

CHAPTER ONE

HELPING HANDS

Humans have hurt Earth's **environment** over the years. When we use **fossil fuels** such as coal, oil, and natural gas, we release greenhouse gases into Earth's atmosphere. Greenhouse gases trap the sun's heat in our atmosphere. This leads to global warming, or an increase in Earth's temperature. Global warming is changing the world's climate.

Caring for the planet is everyone's responsibility.

    Periods of heating and cooling are a natural part of the planet's history. However, scientists agree that Earth is now warming much faster than normal and that human activities are causing this trend. Scientists, engineers, and even kids have been finding ways to tackle many of the problems facing Earth. Thanks to their hard work, there's hope for the future! Read on to learn about new inventions that are helping the planet.

CHAPTER TWO

WHAT IS TECHNOLOGY?

When you think of technology, you probably think of cell phones and smart watches and other things like that. However, technology takes many forms. Technology may be tools, machines, or even a process of doing something.

Our modern idea of technology took shape during the **Industrial Revolution**. During this time, people invented engines to power machines. These machines allowed factories to produce goods at a faster rate. Unfortunately, the engines that powered factory machinery used fossil fuels, especially coal and oil. This was harmful to the environment.

James Watt's steam engine was an important tool during the Industrial Revolution.

As time went on, people began to understand the harmful impact of burning fossil fuels. New technology now makes it possible to produce electricity with cleaner sources, such as solar and wind power. Recent technologies are building on these clean **alternatives**.

CHAPTER THREE

WHAT IS CARBON CAPTURE?

Human activities such as burning fossil fuels and clearing forests have increased the levels of carbon dioxide, a colorless gas, in our atmosphere. This leads to global warming. Scientists are trying to use technology to help solve this problem.

Carbon capture is one of the most promising technologies to help address the carbon dioxide problem. It involves removing carbon dioxide gas from the **emissions** of factories, power plants, and other industrial sites. The gas can be removed in several ways. Once captured, it can be stored underground. It could even be turned into a kind of fuel.

Carbon capture technology can't solve the problem of global warming, but it can help.

In this building in British Columbia, Canada, fans pull in air for carbon capture. The carbon dioxide in the air is **filtered** and removed.

CARBON FILTER

CHAPTER FOUR

NEW AND IMPROVED AIRPLANES

Airplanes produce greenhouse gases, and the number of airplanes in the sky is expected to double by 2040. So, we can only expect more greenhouse gases, right? Wrong! Airplane emissions have been reduced in recent years. Some airplanes are using cleaner **biofuels** rather than jet fuels that pollute the air. Some biofuels create 50 percent fewer greenhouse gases than older jet fuels.

HY4 COCKPIT

The HY4 seats four people. Larger planes are planned for the future.

Some engineers are planning to change airplanes themselves. A plane called the HY4 uses a different kind of motor. It's powered by fuel cells, which create electricity from a mix of hydrogen and oxygen. A battery helps with takeoff. The plane's only emission is water vapor. Researchers plan to study how high and how fast the plane can go. They also want to see how different temperatures affect its performance.

CHAPTER FIVE

TECHNOLOGY AND FOOD

About one-fourth of all land on Earth is used for livestock such as cattle. Sometimes this means the land is cleared of forests, which are needed to absorb, or take in, carbon dioxide. Also, about 14.5 percent of the greenhouse gases caused by human activity are due to raising livestock.

Plant-based foods may help to curb the global warming caused by raising livestock.

New food technology could help solve this problem. Businesses have been perfecting foods that are made from plants but look and taste like meat. One kind of plant-based burger uses pea and rice **proteins**, beet juice, oils, and other ingredients. The company says producing it creates 90 percent less greenhouse gas emissions than a hamburger does. It also claims that if all Americans traded beef burgers for plant-based burgers, it would be like taking 12 million cars off the road for a year.

CHAPTER SIX
SOLAR ENERGY AND SOLAR GLASS

Solar energy is a renewable type of energy. The sun's rays will never run out—at least not for billions of years! Solar energy also doesn't release pollution into the air like the burning of fossil fuels does. Solar cells are the technology used to collect solar energy.

Scientists have invented a new kind of solar technology called solar glass. Solar glass uses tiny solar cells that are smaller than a grain of rice. They're fixed in or on a sheet of glass. Solar glass has been reported to absorb more light than flat solar cells because of its round solar cells. A group at Michigan State University claims that if people installed solar glass in the window space that exists today, about 40 percent of U.S. energy needs could be fulfilled by solar glass.

Scientists are making improvements to solar glass technology.

CHAPTER SEVEN
ENERGY STORAGE

Wind and solar power technologies are promising ways to use renewable energy. However, we need to be able to store the energy these sources create. This can be hard and it can cost a lot of money. Engineers are hard at work coming up with a solution.

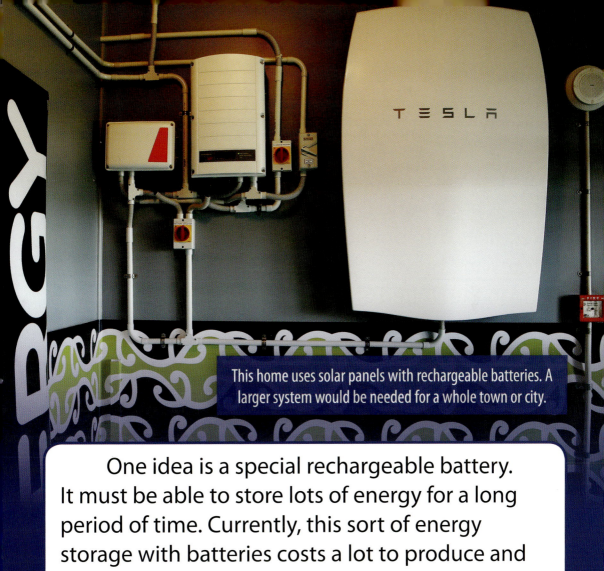

This home uses solar panels with rechargeable batteries. A larger system would be needed for a whole town or city.

One idea is a special rechargeable battery. It must be able to store lots of energy for a long period of time. Currently, this sort of energy storage with batteries costs a lot to produce and maintain, especially for a large area like the United States. In fact, using current methods to store 12 hours of energy for the entire United States would cost $2.5 trillion. So far, it's been challenging to find a battery that's cost effective and can do the job. However, many companies are trying to do just that.

CHAPTER EIGHT
POWER GRIDS

A power grid is a system that brings electricity into our homes. It brings power to the stores and businesses in our towns. The power grid in the United States connects electricity from hundreds of power plants to over 325 million people. However, this grid was designed in the 1890s. It wasn't meant to handle all the energy needs of today. For the most part, the energy in the power grid isn't stored. Only a small part of the system handles renewable energy sources.

In 2018, solar power provided just 1.6 percent of the electrical power in the United States, but it could be much higher in the future.

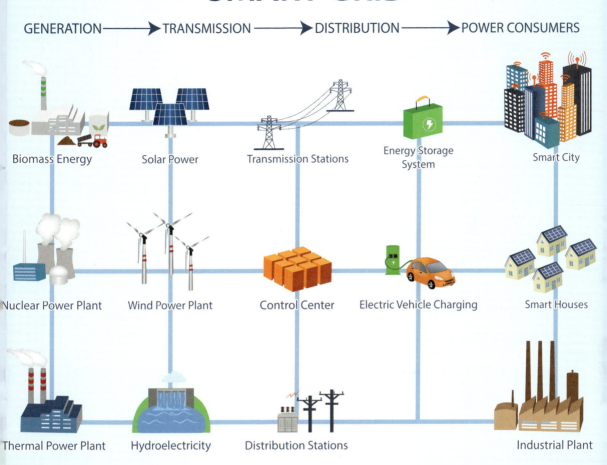

A power grid that uses renewable energy could reduce carbon dioxide emissions by more than 50 percent by 2030. Plans over the next 10 years call for a new kind of power grid called a smart grid. It will be made up of millions of parts.

CHAPTER NINE
FROM PLANT TO PLASTIC

Plastic pollution is another problem facing Earth. Think of all the plastic items in your life, including toys, cups, bottles, bike parts, and more. Plastic is strong and useful, but it can take hundreds of years to break down. Plastic waste builds up over time. It's a major pollution problem that can harm wildlife. Chemicals in plastics can also contaminate, or pollute, soil and water.

More than 90 percent of plastic waste isn't recycled.

In 2014, a young man in Indonesia had an idea. Kevin Kumala began to make plastic from an inexpensive vegetable called cassava. He started a business producing this plant-based plastic. The plastic breaks down after a few months. This makes it great for packaging. The plastic is even safe when **ingested** by animals. Plant-based plastics like this may help fix the plastic problem facing our planet.

CHAPTER TEN

CLEAN OCEAN WATER

Millions of tons of trash end up in the oceans each year. Much of that trash is plastic. Some is dumped on purpose. Some ends up there by accident. Plastic can threaten sea animals' health and may even kill them.

The Seabin is one invention for cleaning up plastic from the ocean. It looks a bit like a floating garbage can. It sucks water, including the trash floating on the surface, in through its top. The plastic collects in a bag. Then the cleaned water returns to the ocean. The Seabin can also collect oil floating on the ocean's surface. Oil can greatly harm ocean creatures and seabirds. More than 850 Seabins have collected over 1 million pounds (453,600 kg) of garbage as of 2020!

Australian surfer and boat builder Pete Ceglinski is one of the inventors of the Seabin.

CHAPTER ELEVEN

CLEAN DRINKING WATER

About 2 billion people around the world don't have a good source of clean drinking water. Good **hygiene** requires clean water for washing. It helps prevent the spread of disease. However, **poverty**, war, or distance from clean water sources makes getting clean water a challenge for many.

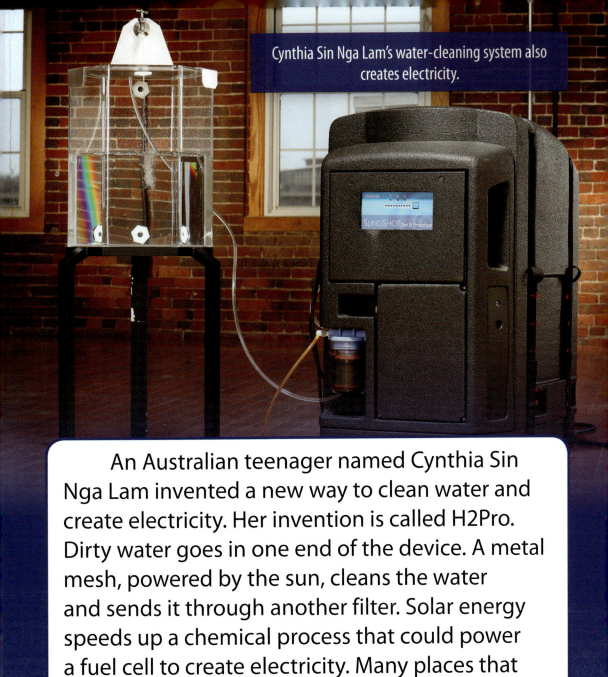

Cynthia Sin Nga Lam's water-cleaning system also creates electricity.

An Australian teenager named Cynthia Sin Nga Lam invented a new way to clean water and create electricity. Her invention is called H2Pro. Dirty water goes in one end of the device. A metal mesh, powered by the sun, cleans the water and sends it through another filter. Solar energy speeds up a chemical process that could power a fuel cell to create electricity. Many places that don't have clean water also don't have steady sources of electricity. Lam's invention could help solve both these problems.

CHAPTER TWELVE

FARMING TECHNOLOGY

Have you ever heard of a vertical farm? Vertical farms are layers of planted crops stacked on top of each other in a kind of a tower. With the help of new technology, vertical farms can be set up in spaces such as old factories or warehouses in cities. That way, the crops don't have to be shipped too far to get to stores.

People can even have small vertical farms in their own homes.

Vertical farming uses no **pesticides** because plants grow indoors. In addition, the crops can't be ruined by storms or extreme weather. The process uses up to 95 percent less water than other forms of farming.

Scientists have found that a mix of red and blue **LED** light helps plants grow best. They're also working on ways to power these farms without using too much electricity.

CHAPTER THIRTEEN

DRONES IN THE FUTURE

Drones are flying vehicles operated by a pilot on the ground. They can be fitted with cameras and sensors and flown to areas that people can't get to easily. Drones can even be used to help the planet.

Pollination is the process that allows plants to reproduce, or create new plants. Bugs such as bees and butterflies normally pollinate plants. Unfortunately, human activities and climate change have put these creatures in danger. The number of bees in particular is dropping.

So, how will crops and flowers be pollinated in the future? Some people are looking for ways to have drones pollinate flowers. Scientists in Japan have been working on small drones that carry pollen from one plant to the next. A test flight has successfully pollinated a flower.

Drones can inspect farms and make sure water and other resources aren't wasted.

CHAPTER FOURTEEN
HELPING YOUR WORLD

We all must take steps to help our planet. We can create goals for ourselves, such as limiting the amount of plastic we use and the trash we create. Technology can help—but we need to help too.

Young people are doing a lot to help the environment. Teens such as Cynthia Sin Nga Lam are leading the way to a brighter future for the planet. You don't have to invent new technology to help.

Activists are people who act strongly in support of a cause. Greta Thunberg is a Swedish teen activist. She gained international recognition for her work. She's calling on world leaders to address climate change.

Stand up for the planet by using your voice. Make changes in your own life. Your actions have the power to create a better future for our world.

GLOSSARY

alternative (ol-TUHR-nuh-tiv) Something that can be chosen instead of something else.

biofuel (BY-oh-fyool) A renewable fuel such as wood, composed of or produced from living matter.

emission (ih-MIH-shuhn) Something that is given off, or the act of producing that thing.

environment (ihn-VIY-ruhn-muht) The natural world around us.

filter (FIL-tuhr) To pass through something to remove unwanted material, or the thing something is passed through.

fossil fuel (FAH-suhl FYOOL) A fuel—such as coal, oil, or natural gas—that is formed in the earth from dead plants or animals.

hygiene (HY-jeen) Ways to keep yourself and your surroundings clean in order to maintain good health.

Industrial Revolution (in-DUH-stree-uhl reh-vuh-LOO-shuhn) An era of social and economic change marked by advances in technology and science.

ingest (ihn-JEST) To take food or liquid into the body.

LED (EHL-EE-DEE) Stands for light-emitting diode, a semiconductor that gives off light when a current runs through it.

pesticide (PEH-stuh-syd) A poison used to kill pests.

poverty (PAH-vuhr-tee) The state of being poor.

protein (PRO-teen) A substance sound in some foods that's very important to the human diet.

INDEX

A
airplanes, 10, 11

B
battery, 11, 17
biofuels, 10

C
carbon capture, 8
carbon dioxide, 8, 12, 19
Ceglinski, Pete, 22
climate, 4, 30

D
drones, 28, 29

E
electricity, 7, 11, 18, 25, 27
emissions, 8, 10, 11, 13, 19

F
fossil fuels, 4, 6, 7, 8, 14

G
global warming, 4, 8, 13
greenhouse gases, 4, 10, 12, 13

I
Industrial Revolution, 6, 7

K
Kumala, Kevin, 21

L
Lam, Cynthia Sin Nga, 25, 30
livestock, 12, 13

P
plastic, 20, 21, 22, 30
power grid, 18, 19

S
Seabin, 22
smart grid, 19
solar energy/power, 7, 14, 16, 18, 25
solar glass, 14

T
Thunberg, Greta, 30

V
vertical farms, 26, 27

W
water, 11, 20, 22, 24, 25, 27, 29
Watt, James, 7
wind power, 7, 16

PRIMARY SOURCE LIST

Page 5
Extinction rebellion protest outside the Bank of England in London. Photograph. Barcroft Media. April 25, 2019. Barcroft Media via Getty Images.

Page 7
James Watt 1788 Double-Action Sun-and-Planet Steam Engine. Photograph. Science & Society Picture Library / Contributor. SSPL via Getty Images.

Page 23
Pete Ceglinski, CEO and co-founder of Seabin Project. Photograph. PETER PARKS/AFP via Getty Images. June 08, 2018. AFP via Getty Images.

WEBSITES

Due to the changing nature of Internet links, PowerKids Press has developed an online list of websites related to the subject of this book. This site is updated regularly. Please use this link to access the list: www.powerkidslinks.com/SOOF/technologies